A BILATERAL DIVISION OF THE PARIETAL BONE IN A CHIMPANZEE; WITH A SPECIAL REFERENCE TO THE OBLIQUE SUTURES IN THE PARIETAL

ALES HRDLICKA

ZINC READ

Published by
This edition copyright © 2024 by Zinc Read
Design and setting By Zinc Read
Email- zincread@gmail.com
ISBN-978-93-6277-216-9

CONTENTS

ARTICLE XXI.--A BILATERAL DIVISION OF THE PARIETAL BONE IN A CHIMPANZEE; WITH SPECIAL REFERENCE TO THE OBLIQUE SUTURES IN THE PARIETAL.

The first to describe a case of division of the parietal bone in apes was Johannes Ranke, in 1899. [1] The skull in question is that of an adolescent female orang, one of 245 orang crania in the Selenka collection in the Munich Anthropological Institute. The abnormal suture divides the right parietal into an upper larger and a lower smaller portion. "The suture runs nearly parallel with the sagittal suture," but, as the illustration shows (Fig. 1), it descends in its posterior extremity towards the temporo-parietal suture, and terminates in this a few millimetres in front of the lambdoid suture. The abnormal suture shows but little serration, and the articulation of the two divisions of the parietal bone is squamous in character, the lower portion overlapping the upper. Below the junction of the abnormal with the coronal suture, the latter takes a pronounced bend forward. A similar bend in the coronal suture is present in the same specimen on the left side. This is common among the other orang skulls in the collection. The portions of the coronal suture below and above the bend differ somewhat in character.

Besides the above-mentioned complete division, Ranke found among the 245 orang skulls 13 with incomplete division of the parietal bone. The division consisted invariably of a longer or shorter remnant of a horizontal "parietal suture," ending in the coronal suture at the top of the bend above referred to. A similar anterior remnant of an abnormal parietal suture was found by Ranke in a young chimpanzee skull; but the author questions the word "chimpanzee," which evidently means that the identity of the skull is somewhat doubtful.

In consequence of his finds, Ranke believes both complete and incomplete divisions in the parietal bone to be much more frequent in the orang than in man. [2] He also thinks that the bend usually present in the coronal suture in the orang signifies that, "even where there are no traces of a parietal suture, such a suture has actually existed in an earlier stage of development." This implies the development of the adult parietal bone in the orang from two original segments, one above the other.

Fig. 1. Division of the Right Parietal in an Orang (Ranke, Abh. d. k. bayer. Akad. d. Wiss., II cl., XX Bd., ii Abth.).

The divisions which I am about to describe occur, one in each parietal, in the skull of a nine-year-old male chimpanzee, which was captured, when young, in West Africa. Later on he was one of the attractions of the Barnum and Bailey Circus, and was familiarly known as Chico. The chimpanzee died in 1894, since when his skin and bones have been preserved in the American Museum of Natural History, New York City. Prof. J. A. Allen, the curator of the Zoological Department of the Museum, has kindly given me permission to describe the skeletal parts for publication. [3]

The most interesting part of Chico is unquestionably the skull. The divisions of the parietal bones which the specimen presents are not only the first complete divisions of the parietal observed in a chimpanzee, but are also unique in character, no divisions of the same nature having been observed before, either in man, in apes, or in monkeys. The position and extent of the divisions in this skull will throw considerable light on the question of the aberrant, complete divisions of the parietal bone, by which term may be designated divisions differing from the typical horizontal ones.

The skull under consideration shows in general a good development and an almost perfect symmetry. The capacity of the brain cavity, measured according to Flower's method, is 390 c.c.

The masculine features of this skull, and particularly the temporal ridges, are not quite as marked as those of another skull of an adolescent male chimpanzee in the Museum. The temporal ridges are slightly prominent, and in their middle third, over part of the frontal and the parietal bones, not more pronounced than in some human crania. They are, however, situated very high. Their upper lines or boundaries touch each other over a part of the sagittal suture, a little back of the bregma; while the lower lines approach to within 6 mm. of the sagittal suture. The supraorbital ridges are not very massive, although prominent to such a degree that, when the skull rests on the occipital condyles and on the teeth, the plane of the orbits is almost vertical. The sagittal crest is insignificant; the occipital crest is high, but not very massive. The zygomatic arches are less strong than they are in an average white male; and the mastoids are small, even smaller than in an average adult white female.

The second dentition is incomplete; the third molars have not reached the level of the opening of their sockets. The condition of the sutures, so far as their patency is concerned, does not bear the same relation to the stage of dentition as it does in man: all the sutures of this skull are more or less obliterated. There are no signs on any part of the skull that point to the closure of any of the sutures as premature. In detail, the condition of the sutures is as follows: The spheno-maxillary articulation is completely closed, but still plainly traceable. Of the various facial sutures, only remnants are open; the suture in the zygomatic arch, however, is almost fully patent on both sides. The spheno-frontal articulation is completely obliterated on the left, but traces of it remain on the right side. The left temporo-sphenoidal and squamo-frontal sutures (the squama of the temporal articulates with the frontal bone) are, with the exception of the basal part of the former, which remains open, quite obliterated, but on the right side both are open. The temporo-parietal sutures, with the exception of 8 mm. of the anterior end of the suture on the right side, are both entirely closed and hardly traceable.

The coronal suture is partly open on the left, and wholly open on the right, up to a point a little below the middle of the anterior border of the parietal bone. At this point on each side, the lower portion of the coronal suture bends backward and continues as the anomalous suture; the upper portion of the coronal, particularly on the right, is completely obliterated, though still traceable. There are no signs left of the sagittal and lambdoid sutures, and only the basal portions of the temporo-occipital articulation remain. The palatine sutures, also, are entirely obliterated.

The skull shows no important anomalies besides the division of the parietals.

The divisions of the parietal bones begin on the left 32 mm., on the right 28 mm. (measured with a tape), above the point of junction of the coronal and temporo-parietal sutures. From the point where the anomalous sutures leave the coronal suture, to the bregma, the distance on the left is 44 mm., on the right 42 mm. The excess of size of the left over the right parietal bone along the coronal suture (6 mm.) compensates the greater height of that portion of the right temporal squama which articulates with the frontal bone. Measured across their middle from the temporo-parietal suture, the two parietals appear to be almost of equal size (left 82 mm., right 80 mm.). In an antero-posterior direction, from the beginning of the division to the middle of the parietal portion of the occipital crest, both bones measure the same, namely 75 mm.

The division in the left parietal begins at a V-shaped cleft, which is filled with a process of the frontal bone. There are slightly distinct markings on the bone and a number of insular ossicles, which make it probable that the cleft had been originally much greater and was largely filled by a Wormian or, rather, a fontanel bone, the lower border of which has subsequently united with the parietal.

For 30 mm. from its beginning the abnormal suture proceeds directly backward, and to this extent shows but little obliteration. The original cleft has, it seems, extended up to this point. From here the suture takes a slight bend upwards, and proceeds almost directly upwards and backwards, becoming gradually obliterated, until it disappears at the temporal ridge, 16 mm. from the median line. Originally the suture must have terminated on the posterior border of the parietal bone, not far from the lambda. The whole suture shows fairly good serration. The coronal suture on this side, below the division, shows serration about equal to that of the abnormal suture; the obliterated portion above this was, so far as can be seen, more simple.

* * * * *

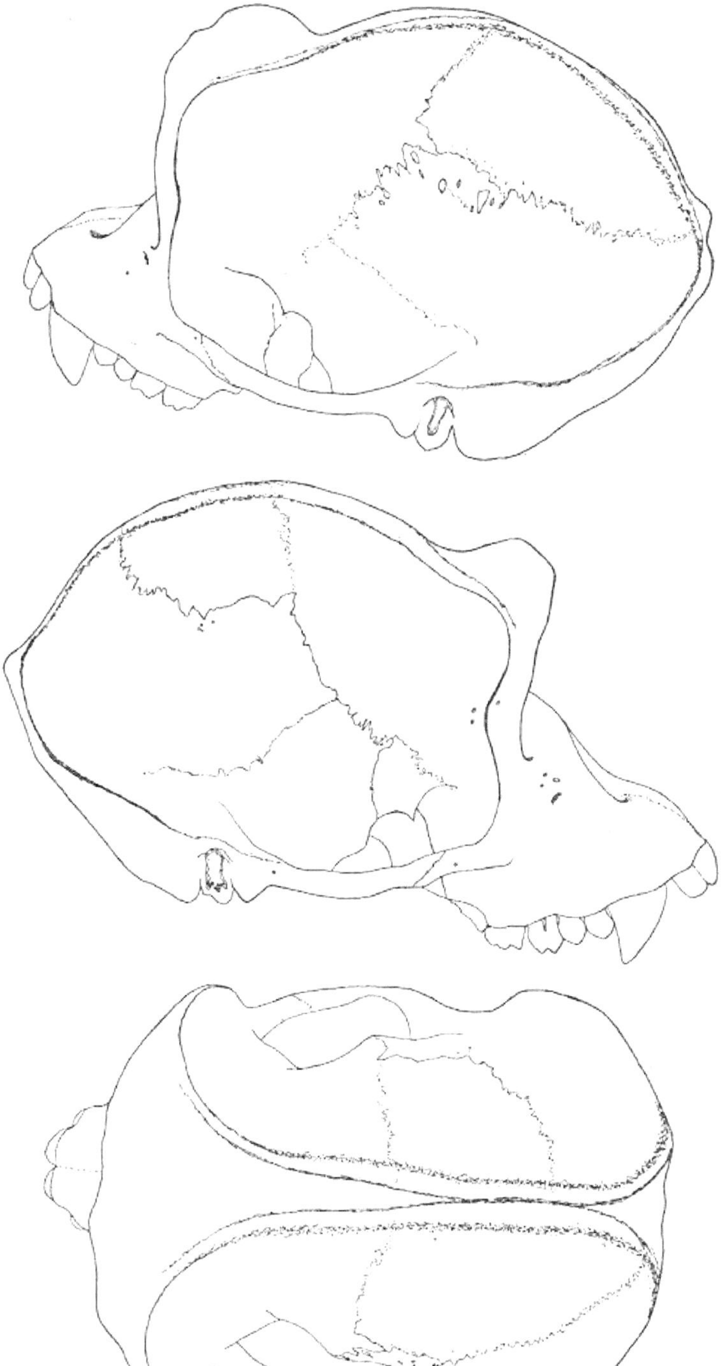

Figs. 2-4. Skull of an Adolescent Male Chimpanzee.

* * * * *

On the right side the division of the parietal may also have begun with a cleft in the anterior border of the bone, but, owing to the advanced state of obliteration of the upper portion of the coronal suture on this side, the existence of the cleft cannot be fully ascertained. Here also the abnormal suture, at first wholly open, runs for the first 26 mm. directly backwards; at this point the suture, still quite patent, takes a turn somewhat sharper than that on the left, and proceeds for 16 mm. backwards and upwards; here it takes a second turn, and proceeds almost directly upwards towards the sagittal suture. This last portion of the abnormal suture is considerably obliterated, and on and beyond the temporal ridge is scarcely traceable. The point at which the division has reached the sagittal suture is situated a little behind the middle of the latter. The abnormal as well as the open part of the coronal suture on this side shows a simpler serration than the corresponding sutures on the left side.

In this specimen there is on neither side any encroachment of the lower portion of the parietal bone upon the frontal, such as Ranke lays stress on in the case of his orangs. A second skull of an adolescent male chimpanzee, in the Museum of Natural History, has a decided bend in the coronal suture, not unlike that which Ranke describes, and which, as he thinks, generally indicates an old parietal division; but in this case the bend is situated between the inferior and superior boundaries of the prominent temporal ridge, and apparently owes its origin to the latter (Figs. 2, 3, 4).

The main interest in the case just described centres in the direction of the abnormal sutures, and in the clearness with which the two divisions appear as equivalent and of the same origin, although one divides the parietal completely, while the other is restricted to one of its angles.

As to the course of the abnormal suture in the parietal bone, in all the cases thus far reported, the division runs in a horizontal direction (cases of Tarin, Soemmering, Gruber,

Hyrtl, Welcker, Turner, Putnam, Dorsey, Ranke, and others); or it runs obliquely from or near the middle of the lambdoid suture to some part of the temporo-parietal suture, the sphenoidal angle, or the lower portion of the coronal suture (cases of Curnow, Ekmark, Gruber, Hyrtl, Lucae, Welcker, Putnam, Traquair, Ranke); in a case of *Simia silenus* described by Gruber and in an Egyptian cranium described by Smith, the divisions run to the lambda and begin respectively slightly above the pterion and at it. In Boyd's and in two of Hyrtl's cases, the abnormal suture begins at or below the bregma on the coronal margin of the parietal bone, and ends at or near its mastoid angle; finally, in Blumenbach's (cited by Welcker), Bianchi's, Fusari's, and Coraini's cases (those of Coraini include two monkeys) the division is vertical, passing between the temporo-parietal and sagittal sutures. The left division in our chimpanzee approaches those in Gruber's *Simia silenus* and Smith's cases; but it originates much higher anteriorly, and terminates slightly below the lambda on the occipital border of the parietal. The division in the right parietal of the chimpanzee, beginning slightly below the middle of the anterior border of the bone, and ending slightly back of the middle of its sagittal border, has no analogy among the cases previously described.

The difference in extent and terminations of the two abnormal sutures in the chimpanzee is of particular interest in connection with the problem of the significance and origin of those divisions of the parietal bone that involve more or less only one of its angles.

Since the observations of Toldt, [4] and more recently of Ranke, [5] on the development of the parietal bone in the human embryo, it appears, though it cannot as yet be said whether the fact is or is not general, that the bone originates from two centres of ossification. These centres appear in most cases one directly above the other, but, as Ranke himself shows, [6] and as can hardly be otherwise, these primitive components of the parietal do not always show the same relations in size or position. The centres blend together, ordinarily, at the end of the third or during the first half of the fourth month of foetal life. On this account, the typical, complete, horizontal division of the human parietal bone, when met with at any time after the fourth month of foetal life, is generally interpreted to-day as a retardation of the union, or a persistence of separation, of the two original segments of the bone. Opinion, however, is still unsettled as to the significance of the more atypical, oblique divisions of the parietal, particularly of those where the separation is limited to one angle. Up to the recent contribution on the subject by Ranke, the weight of opinion on the point, although rather briefly expressed, seems to have been in favor of attributing to these smaller, oblique divisions, the same significance as was given to the more typical, horizontal ones. Gruber, [7] in reporting a new case of a bilateral oblique suture in the parietal bone, calls the separated mastoid angles "the secondary posterior parietals."

Hyrtl and Welcker advance no definite theories on this point, though the latter expresses an opinion [8] that in both the horizontal division and the separation of the mastoid angle of the parietal bone the development of the condition may be identical. In 1883 Prof. F. W. Putnam, in describing one of his Tennessee skulls with an abnormal oblique suture in each parietal, [9] referred the development of the separated mastoid angle on the right side, as well as the larger oblique inferior portion of the parietal on the left side, to a "separate centre" of ossification. Ranke [10] opposes both Gruber's and Putnam's opinion, and presents instead a theory somewhat vague and not satisfactorily demonstrated, by which he accounts for the origin of oblique sutures from partial horizontal sutures in the parietal bone through "half-pathological processes." In his words, "the oblique parietal suture is allied to the half-pathological conditions of the skull; it is wholly unjustifiable to speak, as W. Gruber has done, of a separate Parietale secundarium posterius, severed by the suture, as of a typical, in a certain sense normal, formation. The oblique parietal suture is nothing more than an incomplete (posterior), true, *i. e.*, typical, parietal suture with a sagittal course, modified by certain half-pathological conditions." These half-pathological conditions are produced, the author explains on the preceding page, "durch Einknickung der nach Herrn G. H. Meyer 'plastisch' aufwarts gebogenen hinteren Scheitelbeinrander."

This opinion of Ranke calls for a few words about the incomplete horizontal parietal sutures. These sutures are apparently very rare in human adults, only five instances being on record (4 Ranke's, 1 Turner's). They are more frequent in orangs (Ranke), and quite common (as Ranke shows, and as I found independently before Ranke's publication of his observations) in the human embryos near term and in new-born or very young infants. In the human family, these partial divisions of the parietal generally begin in the posterior part, and run sagittally to the posterior border of the bone, ending in this border at or near its middle. In orangs the incomplete horizontal divisions seem to begin, as a rule, in the anterior part, and end at or near the middle of the anterior border of the parietal. The length of these divisions varies from a few millimetres to several centimetres, and they even reach up to the centre of the parietal bone. [11] These divisions are, without doubt, the remains of the original anterior and posterior clefts, or, if we go a step further, of the original intervening antero-posterior space between the original inferior and superior segments of the parietal. From the very first contact of the growing centres, the median extremity of these clefts is bounded both below and above by a mass of bone; and when the anterior or posterior border of the parietal comes finally in contact with the frontal or occipital bone, the anterior and posterior sagittal clefts, if they still exist, lie between two well-developed, firm portions of the bone. Under these circumstances it is quite impossible to imagine any disturbance, mechanical or pathological, that could affect solely or mainly the median portion of the cleft, and cause a deflection downward in this portion of the division, or cause its extension to the inferior border or even the anterior-inferior angle of the parietal.

There are only two factors that can possibly affect and modify the course of the incomplete parietal suture, and both of these would show their influence mainly or entirely on the distal portion of the same. These two factors are, first, an abnormal development, either defective or excessive, of one of the original parietal segments; and, secondly, influences that would interfere with the freedom of full growth of the anterior or posterior border of the parietal.

In the first case, as can easily be imagined or even artificially demonstrated, there would be possible only a lower or higher situation or an obliquity affecting mostly the marginal portion of the division. The results would be low or high sagittal sutures, and curved or oblique sutures diverging from the parietal eminence,--effects entirely different from the actually observed oblique sutures that sever the lower portion of the parietal, or its mastoid angle.

Influences interfering with the free development of the anterior or posterior border of the parietal bone could only deflect upwards or downwards the marginal end of an incomplete parietal suture, or, at most, in a case of a short suture, render it oblique or curved in its entirety. No pathological condition, unless it were accompanied by a fracture, could extend even a deflected antero-posterior incomplete division to any of the borders of the bone.

There are, it seems to me, only three possible ways in which an oblique suture, extending between any two borders of the parietal bone, can be produced.

In the first case the oblique suture, or rather a suture-like formation, may be the effect of an early fracture. A fracture produced in adult life is generally recognizable as such; but a fracture dating from earlier stages of life, produced before the growth of the bone has ceased, may, if not entirely obliterated, present more or less the characteristics of a suture. I have seen several skulls where a division in the parietal bone or the temporal squama presented at the same time features of a fracture and suture; in one or two of these cases so much so, that it was and still is impossible for me to decide exactly which of the two conditions I had before me. Gruber describes one such case [12] as an instance of an oblique parietal suture, while Hyrtl and Ranke both consider this case as one with an acquired division. To differentiate a congenital real oblique suture from a division which is the result of a fracture, we must be guided largely by the situation, form, and serration of the division, and the condition of the surrounding bones, especially that of the opposite parietal. A straight course, ending with one extremity in or near the middle of the anterior or posterior border of the parietal, a complex serration, no continuity of the division on the neighboring bones, and particularly a coexistence of an allied or similar division on the opposite parietal,--all favor the conclusion that the division under consideration is a real congenital suture, and not the result of a fracture.

In the second case there are reasons for believing that an oblique suture of the parietal bone can originate in the same way as the horizontal one, namely, through a persistence of the original separation between the two centres from which the bone is developed, and a co-existent difference in the relative position or the relative growth of the two centres. It is in this connection that the above-described division in the parietals of the chimpanzee will prove of value.

The occasional persistence of the separation between the two original segments of the parietal bone is sufficiently demonstrated by the presence of the complete horizontal parietal suture. Differences in the relative position of these segments can be observed in a limited degree in Ranke's illustrations of embryos, before referred to; it can be deduced from such cases as the two of Hyrtl, [13] in which the division of the parietal was directed from the upper portion of the anterior to the lower portion of the posterior border of the bone. The most pronounced change in the position of these centres may be witnessed in cases where the parietal bone shows a perfect vertical instead of a horizontal suture. Such cases have been referred to before, and I presented at the meeting of the Association of American Anatomists, in 1899, several such examples, found by me in skulls of monkeys in Professor Huntington's anatomical collection in the Medical Department of Columbia University. One of these specimens is shown in the accompanying illustration (Fig. 5).

A difference in the relative growth of the two centres of the parietal bone is well shown in the difference of size between the inferior and superior portions of the parietal in cases of the complete horizontal suture in the same. In the majority of such cases on record the superior portion is larger, particularly anteriorly, than the inferior; so much so, that that condition seems to be the typical one. The difference in the size of the two portions of the parietal, and in their relative anterior and posterior height, is most pronounced in one of Gruber's cases, [14] where the "parietal suture" begins only 10 mm. above the pterion, and ends 40 mm. above the asterion. In Dorsey's case [15] the lower portion of the divided parietal is 12 mm. higher than the upper. The same condition as is found in Gruber's case, here mentioned, exists in the almost identical left division of the second case of Putnam, of which I have a photograph in my hands. A somewhat similar excess of the posterior over the anterior part of the lower severed portion of the parietal can also be seen in the illustrations of the cases of Tarin, Lucae, and Turner (Admiralty Islands skull). In Calori's interesting case [16] there is a decided excess of the lower portion of the divided parietal in its posterior portion on the left and in its anterior portion on the right side.

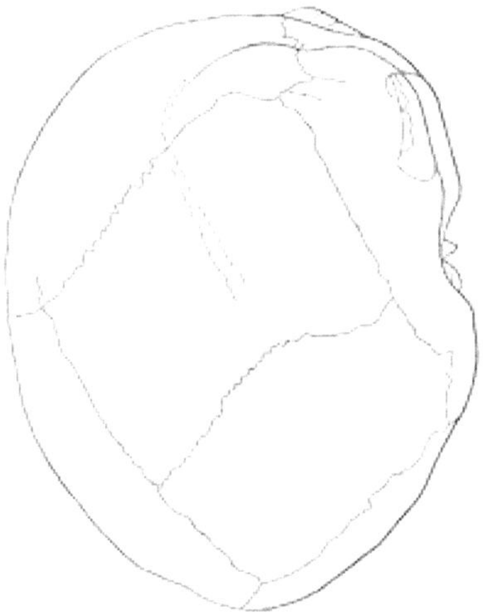

Fig. 5. *Macacus rhesus* (Medical Department, Columbia University), showing a Complete Division of the Right Parietal Bone in a Vertical Direction.

In case the upper segment was not vertically above the lower one, but in a position a little more forward or backward of it; and, furthermore, if the relative growth of the two segments differed, and their separation remained permanent,--the separation of any portion of the parietal bone in almost any form and to almost any extent might result. Such coincidence of anomalous conditions, although necessarily rare, cannot, from what we know on the subject in parietal and other bones, be declared improbable. All cases where oblique suture on one side co-exists with more or less horizontal suture on the other side in the parietal bone, as in the second of Putnam's cases, would of course point directly to a similar origin of the anomaly on both sides of the cranium. That such cases have not been more frequently observed is largely due, I think, to the rarity of bilateral parietal divisions.

A third mode of development of the oblique suture in the parietal bone suggests itself where the severed portion of the bone is small, and that is the possible existence of a supernumerary, third centre of ossification. I am by no means ready to defend this theory, yet there are cases in which it would afford the easiest explanation. I have a Peruvian skull at hand, in which there is a bilateral, quite symmetrical quadrangular separate piece of bone, encroaching on the mastoid process of the parietal. The surface of the left parietal bone in this skull measures across its middle in antero-posterior direction 120 mm., in infero-superior direction 130 mm.; similar measures of the right parietal are respectively 117 and 130 mm. The separate bone on the left measures across its middle in antero-posterior direction 20 mm., in infero-superior direction 12 to 21 mm.; the same portion on the right measures respectively 25 and 11 to 15 mm. Both pieces are joined to the parietal bone by a squamous suture (Fig. 6).

It is apparent that the separate pieces of bone in this case are too small to be easily taken for representatives of one of the regular centres of ossification of the parietal bone; but the same pieces are somewhat too large, and especially too singularly outlined and joined to the parietal, to be without difficulty diagnosed as simple Wormian or fontanel bones. One of Ranke's cases, [17] though the separation of the mastoid angle is oblong instead of quadrangular, as in the Peruvian skull, seems to me to present a similar difficulty in properly diagnosing the nature of the severed portion. This group of cases needs further observation, particularly on the bones of infants and embryos. I have two monkey skulls at hand which actually show a multiplicity of the original segments of the parietal. These specimens will be described in a future publication.

Fig. 6 (99/3550). Quadrilateral Fontanel Bones in a Peruvian Male Skull, encroaching upon the Mastoid Angle of the Parietals.

So much as to the *formation* of the oblique sutures in the parietal. It should not be forgotten that such sutures can be *simulated* by those which divide true Wormian or fontanel bones from the parietal. The distinction between the real oblique parietal and these extra-parietal sutures must depend largely on the extent of the division and form of the separate piece of bone.

We may now return to the skull of our chimpanzee. In considering the nature of the divisions in the parietal bones of this skull, we can at once and absolutely discard the idea of the divisions being due to fractures, or being boundaries of Wormian or fontanel bones, and thus really extra-parietal in their nature. There is nothing about the sutures, or the divided pieces, or the neighboring bones, that would even suggest such an explanation; and in our records on Wormian and fontanel bones we find no analogies either in man, or apes, or lower animals, to the conditions here observed. The necessary conclusion from this can only be that we have before us two examples of real parietal division.

The division on the left side, had it existed alone, would be readily acceptable as an instance of the "parietal suture." The anterior extremity and more than the anterior third of the course of the division correspond exactly to the same features of a typical, horizontal "parietal suture;" while the elevation of the posterior extremity of the division, though unusual, can readily be explained as due to an excess in growth of the inferior original centre of the bone, which may, in addition, have been situated slightly posterior to the upper centre.

The division in the right parietal of the chimpanzee begins at its anterior end, and runs for the first third of its course in the same way as that on the left side; its posterior end, however, does not reach the lambdoid, but turns up and ends in the sagittal border. Should this formation have existed alone, I should be inclined to consider it either as the result of an accessory centre of the parietal, or, possibly, as a persistence of the anterior portion of the divided superior centre of the bone, the posterior portion of the same being united with the lower segment of the parietal in the usual way. With the division of the left parietal in the same skull before me, everything points to a similar origin of the division on both sides, and to the right as well as the left division being a true "parietal suture," deflected less on the left and more on the right side by a disproportion in growth of the two original, regular segments of each of the bones.

The disproportion of growth of the two original segments of the parietal bone will, I believe, be found more common as attention is directed to this subject. It can be well explained, though there may at times be other factors present, by a difference in the blood-supply to the two centres. This of course may occur not only in different skulls, but also on the two sides of the same cranium.

THE END